Ageing

By

Dr. Steve J. Hayes

Table Of Contents

Introduction

Early adulthood is the beginning of a long, ongoing process of natural change called aging. Many body processes start to gradually deteriorate in the early middle years. At no certain age do people become old or elderly. Old age has traditionally been defined as commencing at age 65.

Chapter 1

Taurine

Taurine is an amino corrosive that plays a couple of significant parts in your body, including supporting safe well-being and sensory system capability. More often than not, your body creates sufficient taurine all alone, however, enhancements can likewise assist you with meeting your taurine necessities.

On one occasion after secondary school, I halted at a general store with my companion to get a caffeinated drink before our exercise.

Filtering the coolers of ready-to-go beverages, my companion called attention to Red Bull. He guided me to the combating bulls on the can and made sense of that Red Bull contained a fixing called taurine, which he asserted was separated from bull semen.

Shocked and confounded, I chose to go with a contending brand to empower my exercise that day.

It was only after my initial school years, when I turned out to be keener on nourishment and sports supplement research, that I took in my companion's case wasn't precisely exact.

Today, Red Bull keeps on remembering taurine for its recipe. You can likewise track down taurine in an assortment of pre-exercise and energy supplements. Further, a few food sources contain it, and your body might in fact create it.

This article makes sense of all that you want to be aware of taurine, including its advantages, its incidental effects, and whether and how you ought to take taurine enhancements.

What is taurine?

Taurine is a normally happening sulfur-containing amino corrosive. It's especially gathered in your cerebrum, eyes, heart, and muscles.

Albeit amino acids are frequently alluded to as the structure blocks of protein, taurine isn't utilized to construct proteins in your body. All things considered, it's viewed as a restrictively fundamental amino corrosive, meaning it becomes fundamental just in the midst of sickness and stress.

Have confidence that, in spite of the normal conviction, taurine isn't removed from bull semen or pee. Rather, it was first segregated in 1827 from the bile of a bull. The Latin name for a bull is Bos taurus, which is where the amino corrosive's name started.

Taurine is tracked down in certain food varieties, and your body might in fact create it without anyone else. Along these lines, taurine lack is improbable in sound grown-ups.

Nonetheless, on the grounds that babies and newborn children can't make taurine as well as grown-ups, they rely upon taurine from bosom milk or taurine-enhanced recipe.

Outline

Taurine is an amino corrosive tracked down in specific food varieties. Your body can likewise make it. It's fundamental just in specific conditions, like in the midst of sickness and stress.

Wellsprings of taurine

The fundamental wellsprings of taurine are creature proteins like meat, fish, and dairy. Plants contain no obvious measure of taurine.

Thus, individuals eating a veggie lover or vegan diet consume less taurine. They will generally have lower taurine levels than the individuals who routinely eat creature proteins.

All things being equal, taurine inadequacy is improbable. This is thanks to your body's capacity to make taurine in your liver from other amino acids.

As well as getting taurine from food, you can get it from some caffeinated drinks. These normally give around 750 mg for each 8-ounce (237-mL) serving.

For reference, the normal American eating routine gives 123-178 mg of taurine day to day, while a lacto-ovo vegan diet — which incorporates both dairy items and eggs — gives something like 17 mg of taurine every day.

The type of taurine utilized in enhancements and caffeinated drinks is typically manufactured, meaning it's not got from creatures. In this way, it's reasonable for those eating a veggie lover or vegan diet.

Synopsis

The vital dietary wellsprings of taurine are protein-rich creatures and food sources like meat, fish, and dairy. Taurine is tracked down in more modest sums in some plant food sources. It's likewise added to numerous caffeinated drinks.

Capabilities in your body

Taurine is tracked down in a few organs and has far and wide advantages.

The principal jobs of taurine in your body are:

- keeping up with appropriate hydration and electrolyte balance in your cells

- shaping bile salts, which assume a significant part in assimilation

- managing minerals like calcium inside your cells

- supporting the general capability of your focal sensory system and eyes

- managing invulnerable framework well-being and cell reinforcement capability

Since taurine is a restrictively fundamental amino corrosive, a sound grown-up's body can create the negligible sum expected for these fundamental everyday capabilities.

Be that as it may, your body might require bigger sums in the midst of sickness or stress. This might be the situation in individuals with heart or kidney disappointment and in untimely newborn children who have been taken care of intravenously. These people might have to get taurine from food or enhancements.

In creature models, taurine lack has been displayed to cause eye harm, ongoing liver sickness, muscle-debilitating, and an expanded gamble of creating diabetes.

Taurine lack of people is uncommon, so its belongings remain generally obscure. In any case, low taurine levels have also been related to these circumstances.

Rundown

Taurine assumes numerous significant parts of your body. Albeit very uncommon, taurine lack has been displayed to prompt serious medical problems in creature studies.

Benefits

Due to its overflow in the body, its cancer prevention agent and mitigating properties, and its job in energy creation, taurine has been read up for its possible job in overseeing different clinical circumstances and further developing activity execution.

May battle diabetes

Taurine's cancer prevention agent and mitigating properties might upgrade insulin awareness, subsequently diminishing the gamble of type 2 diabetes or further developing glucose the board in those with the condition.

For sure, one investigation discovered that individuals with diabetes have a 25% lower centralization of taurine than those without diabetes. This proposes that taurine might play a part in diabetes the board.

Albeit momentum research on the impacts of taurine enhancements for diabetes the executives in people are restricted, a 2018 survey proposes that the enhancements could be a decent restorative choice for further developing glucose the board in individuals with diabetes.

A similar survey likewise proposes that taurine could have defensive impacts against diabetes-related confusions, for example, nerve harm, kidney harm, and coronary illness.

In any case, it's obscure whether low taurine levels are a reason or a result of diabetes, and more examination is required.

May further develop heart wellbeing

Taurine enhancements have been displayed to control circulatory strain and further develop heart capability and blood fat levels in individuals with heart conditions like a cardiovascular breakdown. At significant levels, it might try and safeguard against coronary illness.

Research recommends a connection between higher taurine levels and decreased cholesterol, lower circulatory strain levels, and fundamentally lower paces of death from coronary illness.

In one review, individuals with cardiovascular breakdown took 500 mg of taurine multiple times day to day for quite a long time.

They encountered huge decreases in degrees of all-out cholesterol, fatty substances, and C-responsive protein (CRP) — a provocative biomarker — both when working out, contrasted and the people who took a fake treatment.

In a 12-week concentrate on individuals with a high-typical pulse, requiring 1.6 grams of taurine each day decreased systolic circulatory strain (the top number) by

7.2 mmHg and diastolic circulatory strain (the base number) by 4.7 mmHg contrasted and fake treatment.

Taurine might assist with lessening hypertension by diminishing the opposition of the bloodstream in your vein walls and by working on the effectiveness of skeletal and heart muscle compressions.

May help practice execution

Due to its capacity to improve muscle constriction and defer muscle weakness, taurine might help athletic execution.

Furthermore, taurine might increment fat consumption during activity to all the more likely fuel your presentation.

A survey of 19 examinations evaluating the impacts of taurine on athletic execution noticed a few advantages, including:

- expanded oxygen take-up by the body

- expanded chance of weariness

- diminished muscle harm

- further developed recuperation times

- further developed strength and power

The survey creators propose that a viable portion to accomplish these advantages is 1-3 grams required 1-3 hours before your exercise for somewhere around 6-21 days.

In any case, the creators likewise note that taurine's impacts on practice execution will more often than not belittle and conflict. In this manner, more exploration is required on the point.

Other medical advantages

Other likely advantages of taking taurine enhancements incorporate:

- May help eye well-being. Taurine's cell reinforcement impacts might assist with combatting the oxidative pressure related to retinal degenerative illnesses, for example, age-related macular degeneration.

• May help to hear. Taurine might forestall the hair cells inside the ear from becoming harmed, which is a vital supporter of hearing misfortune.

• May offer neuroprotective impacts. The mitigating impacts of taurine might diminish irritation inside the mind and battle neurodegenerative circumstances like Alzheimer's sickness.

• May uphold liver well-being. Taurine might have defensive impacts against persistent and intense liver injury.

Albeit promising, these potential advantages are less examined or are principally upheld by creature and test-tube reviews. Consequently, more examination is expected to look into taurine's advantages for human well-being.

Outline

Taurine might help individuals with diabetes, further develop coronary illness risk factors, and improve different parts of athletic execution. It might likewise offer an extensive variety of other potential medical advantages, however, it is missing to help prove.

Aftereffects and precautionary measures

As indicated by the most ideal that anyone could hope to find proof, taurine has no bad incidental effects when enhanced suitably.

One 2019 report proposes that the most elevated day-to-day portion of taurine you can securely consume is 3 grams each day. Notwithstanding, the European Sanitation Authority (EFSA) proposed in its 2012 rules that you can securely take as much as 6 grams each day.

All things considered, certain individuals have detailed incidental effects subsequent to taking taurine, including:

- spewing

- queasiness

- liver torment

- migraine

- stomach torment

It's hazy whether these secondary effects are connected with the amino corrosive or to an alternate fixing that might have been taken close by taurine.

It's quite important that albeit no proof shows that taking taurine close by physician-endorsed drugs causes incidental effects, it goes about as a cytochrome P450 protein inhibitor.

This implies it could disrupt meds that depend on this compound to use drugs, for example, antidepressants, antiepileptic medications, warfarin, and statins.

Accordingly, assuming that you're utilizing any drugs, you ought to counsel your primary care physician to see if there's any gambling related to taking taurine.

Additionally, assuming you decide to expand your taurine admission through pre-exercise enhancements or caffeinated drinks, consider some other fixings in these items that you might be delicate to or need to restrict. For instance, these items might be high in caffeine or added sugar.

Synopsis

At the point when consumed in sensible sums by a sound individual, taurine makes no known side impacts. All things considered, it might connect with specific medications, so counsel your primary care physician prior to taking taurine on the off chance that you're taking any meds.

Step-by-step instructions to enhance

The most widely recognized measurement range for taurine is 500-3,000 mg each day.

In any case, remember that an EFSA report from 2012 recommends that up to 6,000 every day is protected, showing major areas of strength for its profile.

While certain examinations might involve a higher portion for brief periods, adhering to 3,000 mg each day will assist you with boosting the advantages while remaining inside a protected reach.

The least demanding and most practical method for arriving at this measurement is through powder or case supplements. Most case supplements contain 500-1,000 mg for every serving, while powdered taurine can have 1,000-2,000 mg for each serving.

I would say, taurine powder blended in with water has a somewhat unpleasant taste, so you might need to try different things with various blenders to find a flavor profile you like.

Rundown

Enhancing with 500-3,000 mg of taurine each day is known to be powerful and safe.

The reality

Taurine is a restrictively fundamental amino corrosive, meaning you really want a greater amount of it during times of pressure or sickness. Lack is for the most part uncommon on the grounds that taurine is normal in creature protein food varieties and in light of the fact that your body can make it in your liver.

Taurine enhancements have been read up for their restorative jobs in overseeing diabetes and coronary illness risk factors, for example, hypertension. They likewise show a guarantee for working on different proportions of sports execution. In any case, more exploration of people is required.

Taurine has areas of strength for a profile, however, remember that it might interface with specific prescriptions, so it's ideal to converse with a medical care proficient prior to taking it.

Chapter 2

Tauring aging

Taurine Safeguards Maturing Minds

An amino corrosive called taurine spikes the development of new synapses, even in advanced age. Taurine has likewise been displayed to help safeguard against factors associated with neurological decay and stroke harm.

For a really long time, it was expected that new synapses quit being created once an individual is completely developed.

Late discoveries demonstrate that isn't accurate. Grown-ups at whatever stage in life have the capacity to make new synapses.

A minimal-expense supplement called taurine has been displayed to improve this young reestablishing process.

Taurine safeguards against neurotoxins in a manner that can slow progress in years-related neurological decline.

Preclinical investigations show that taurine lessens the sort of harm that individuals experience because of stroke or injury to the brain.

The higher admission of taurine prior to life is related to better mental capability in the elderly.

The body makes just modest quantities of taurine.

To support taurine levels, it should be obtained from the eating routine or through supplementation.

What You Want To Be aware

- Taurine is basic for typical mind capability.

- It keeps up with a solid mitochondrial energy supply, which is normally reduced with propelling age.

- It spikes the development of new synapses, even in advanced age.

- Keeping up with sufficient taurine levels might assist with lessening age-related dangers to mental capabilities, like neurotoxins, oxidative pressure, and aggravation.

- In preclinical examinations, taurine has been displayed to help cerebrum capability while protecting against dementia-related changes and different types of mind injury.

The Elements of Taurine

Taurine is an amino corrosive that keeps up with solid cells all through the body.

It is basic for the pinnacle working of mitochondria, the "power plants" that supply cells with energy.

Mitochondrial capability melts away with age, debilitating cell capability and making it harder for cells to safeguard themselves from stress and injury.

This is particularly obvious in the brain, quite possibly the most metabolically dynamic organ in the body.

Taurine additionally has numerous alternate ways it might help safeguard and advance cerebrum wellbeing, including:

- Animating new synapse growth,

- Safeguarding against excitotoxicity (unsafe overactivity in the brain),

- Safeguarding against poisons that can harm the brain,

- Limiting mind harm brought about by stroke and head wounds, and

- Safeguarding typical cell capability and energy supply by controlling calcium, safeguarding cell films, and more.

Through these instruments, taurine might assist with safeguarding the maturing mind against mental weakness, dementia, and harm from strokes, head wounds, and neurotoxins.

Neurogenesis: Keeping the Cerebrum Adjusted

Preclinical examinations show that taurine lifts the making of new mind cells. This is called neurogenesis, and it is a vital aspect of safeguarding sound cerebrum capability as we age.

At the point when neurogenesis happens during mental health, it alludes to synapses developing, partitioning, and developing. Yet, neurogenesis doesn't stop after improvement.

Over the course of life, we should keep up with solid, useful synapses. We likewise need to safeguard the associations between those synapses. All our mind capabilities — from controlling development to undeniable level mental assignments like talking, learning, and recollecting — require these fundamental associations.

Neurogenesis decreases with age. Therefore, our minds contract. The basic brain associations then become lessened. This can begin the elusive slant into mental disability and dementia.

By empowering the formation of new cerebrum cells,1 taurine could assist with combatting age-related decrease in mental capability.

A few cell and creature concentrates on the show that taurine makes a difference in "awakening" the mind of immature microorganisms, invigorating new synapse creation, and supporting their survival.

This kind of impact has been seen in both youthful, creating cerebrums and in more seasoned minds. In one investigation of moderately aged mice, taurine enacted stem cells.1 Fundamentally, this occurred in the hippocampus, a mind locale basically significant for the development of new recollections. This has clear repercussions in the battle against Alzheimer's and dementia.

As well as actuating foundational microorganisms, taurine delivered new cells and upheld their endurance, while additionally decreasing unsafe aggravation in the brain.

Guard Against Excitotoxicity

Excitotoxicity has for some time been known to add to the harm that happens after awful cerebrum injury and stroke.27 All the more as of late it has likewise been connected to the movement of dementia in the elderly.28

Excitotoxicity is the cycle by which synapses are harmed or killed when certain receptors in the cerebrum are overactivated. This is most frequently seen with the synapse glutamate.

Glutamate is basic for typical mind capability. It conveys messages between nerve cells and assumes a part in learning and memory.

Overstimulation of synapses by persistent, elevated degrees of glutamate causes brokenness and customized cell death.

Taurine safeguards against excitotoxicity by impeding and decreasing the overstimulation brought about by the abundance of glutamate. Typically, cells kick the bucket not long after they're presented with high convergences of glutamate. In any case, when pre-treated with taurine, cells make due under these circumstances.

This is a significant finding given how significant excitotoxicity is in the improvement of numerous normal cerebrum problems.

Cerebrum Harming Beta-Amyloid

Taurine's capacity to safeguard against excitotoxicity and end cell passing makes it undeniably fit to forestall age-related degeneration of the sensory system, including Alzheimer's and dementia.

Creature concentrates on showing that taurine likewise lessens 2 extra factors that add to mental degradation and a chance for dementia: the harmful impacts of beta-amyloid gathering and inflammation.

In a review distributed in the diary Neuropharmacology, taurine safeguarded rodent synapses from the harmfulness regularly prompted by beta-amyloid deposits.

In any event, a portion of this defensive impact gives off an impression of being because of taurine's capacity to tie to beta-amyloid straightforwardly. A review utilizing a mouse model of Alzheimer's illness saw this amyloid-restricting, which corresponded with enhancements in mental capability on different tests.

A rodent model of mental degradation likewise showed that taurine safeguarded cerebrum capability, protecting against oxidative pressure, helping neurotransmission, and diminishing mind inflammation.

Assurance Against Dementia

This multitude of studies showing taurine's defensive impacts proposes that supplementation with it very well may be a significant preventive measure against Alzheimer's and dementia.

A gathering of specialists in South Korea investigated the connection between taurine and security from dementia in the elderly. They assessed the previous admission of taurine in 40 more established people with dementia and contrasted it and that of 37 sound individuals of a similar age.

What they found was that older individuals with dementia had a fundamentally lower level of taurine admission than solid subjects did when they were more youthful. The typical admission of taurine in solid subjects was roughly 18% higher than in the people who created dementia.

This concentration additionally showed how much taurine admission corresponded with the level of mental capability. As such, the individuals who had the most noteworthy admission of taurine had the best scores on mental tests.

In another review, on older ladies, 1,500 mg of taurine every day diminished aggravation, safeguard the wellbeing of the blood-mind boundary, and further develop mental grades north of 14 weeks.

These examinations demonstrate that taurine backings sound cerebrum capability and may safeguard against Alzheimer's illness and dementia.

Reducing the Effect of Stroke

Strokes can devastatingly affect discernment and cerebrum capability.

They can be for the most part isolated into 2 significant sorts: ischemic stroke and hemorrhagic stroke.34

Ischemic stroke is more normal and happens when the bloodstream is decreased or obstructed to a piece of the mind, prompting cell demise and deficiency of function.

In a creature model of ischemic stroke, supplemental taurine diminished the volume of mind harm brought about by a stroke by around 55% contrasted with creatures that didn't get treatment.

Moreover, a few markers of injury seriousness, remembering oxidative pressure and energy creation for the cerebrum, were diminished in creatures given supplemental taurine.

Hemorrhagic stroke alludes to abrupt, unconstrained seeping into or around the mind. Albeit more uncommon than ischemic stroke, it actually influences numerous more seasoned people, especially those with high blood pressure.

A rodent model of hemorrhagic stroke showed that taurine safeguards against the mind harm brought about by this sort of stroke as well.8 Rodents given taurine had decreased loss of capability with drain and experienced less cerebrum enlarging and irritation.

Decreasing Harm Because of Head Injury

Head wounds are troublesome for older people. Injury from falls or different mishaps can cause a critical loss of mental capability.

A few preclinical investigations have shown that taurine works on the results of these kinds of injuries.

In one review, taurine forestalled synapse harm after trial head injury in rats. The treated creatures likewise experienced upgrades in the cerebrum bloodstream and improved mitochondrial capability.

A few other creature studies have shown that taurine shields the mind from harm, yet additionally further develops capability after a head injury.

Decreasing the Impacts of Neurotoxins

Taurine might assist with shielding the cerebrum from neurotoxins that harm the sensory system.

One of the most well-known harmful compounds is glucose, particularly at the significant levels found in diabetes. In creature models of diabetes, raised glucose prompts irritation, oxidative pressure, and DNA harm in the mind.

In a new report, trial diabetes in rodents caused this multitude of destructive cell changes in numerous regions of the brain. However treatment with taurine decreased this multitude of impacts.

Taurine likewise safeguarded the minds of creatures presented with the harmful impacts of a few mixtures, including arsenic, unpredictable gases, and other known neurotoxins.

Rundown

Taurine is an amino corrosive that is basic to solid cell capability, especially in profoundly dynamic tissues like the cerebrum.

Strikingly, taurine seems fit for helping the production of new synapses at whatever stage in life.

Taurine additionally safeguards against harmfulness, oxidative pressure, and aggravation. Accordingly, taurine might assist with forestalling age-related mental deterioration, dementia, and injury from stroke and head injury.

Just modest quantities of taurine are delivered to the body. By enhancing with taurine, you can assist with keeping up with ideal levels essential for dragging out top mental capability into advanced age.

Luckily, taurine is certainly not a cumbersome amino corrosive, implying that 1,000 mg can be gotten by taking only one case every day.

Chapter 3

Accounts receivable aging report

What Is accounts receivable aging report?

Money due maturing is an occasional report that orders an organization's records receivable as per the timeframe a receipt has been remarkable. It is utilized as a check to decide the monetary well-being and dependability of an organization's clients.

In the event that the records receivable maturing shows an organization's receivables are being gathered substantially more leisurely than typical, this is an admonition sign that the business might be dialing back or that the organization is assuming more prominent praise risk in its deals rehearses.

KEY Important points

• Money due maturing is the most common way of recognizing open records receivables in view of the time allotment a receipt has been extraordinary.

• Records of sales maturing are valuable in deciding the recompense for suspicious records.

• The matured receivables report organizes those solicitations owed by length, frequently in 30-day fragments, for speedy reference.

• Records of sales maturing are utilized to assess the worth of receivables that the organization doesn't anticipate gathering.

• This data is utilized to change the organization's fiscal summaries to try not to exaggerate its pay.

How Records Receivable Maturing Functions

Money due maturing, as an administration instrument, can show that specific clients are becoming credit gambles, and may uncover whether the organization ought to continue to work with clients that are persistently late payers.

Records of sales maturing have sections that are ordinarily broken into date scopes of 30 days each and show all out receivables that are right now due, as well as those that are past due for every 30-day time period.

Remittance for Dubious Records

Money due maturing is helpful in deciding the recompense for far-fetched accounts. While assessing how terrible the obligation is to investigate an organization's budget summaries, the records receivable maturing report is valuable to gauge the aggregate sum to be discounted.

The essential valuable element is the accumulation of receivables in light of the period of time the receipt has been past due. Accounts that are over a half-year-old are probably not going to be gathered, besides through assortments or a court judgment.

Organizations apply a decent level of default to each date range. Solicitations that have been past due for longer timeframes are given a higher rate because of expanding default risk and diminishing collectibility. The amount of items from each extraordinary date range gives a gauge regards to the absolute of uncollectible receivables.

The IRS permits organizations to discount matured receivables, however, provided that the organization has abandoned gathering the debt.

Matured Receivables Report

The matured receivables report is a table that gives subtleties of explicit receivables in light old enough. The particular receivables are collected at the lower part of the table to show the all receivables of an organization, in view of the number of days the receipt is past due.

The ordinary segment headers incorporate 30-day windows of time, and the lines address the receivables of every client. Here is an illustration of a records receivable maturing report.

Advantages of Records Receivable aging

The discoveries from records of sales maturing reports might be worked on in different ways. To start with, money due is a determination of the expansion of credit. In the event that an organization encounters trouble gathering accounts, as confirmed by the records receivable maturing report, issue clients might be expected to carry on with work on a money-just premise. In this way, the maturing report is useful in spreading out credit and selling rehearses.

Records of sales maturing reports are additionally expected for discounting awful obligations. Following delinquent records permits the business to gauge the number of records that they can not gather. It likewise assists with recognizing potential credit dangers and income issues.

Organizations will utilize the data on a records receivable maturing report to make assortment letters to ship off clients with late equilibriums. Debt claims maturing reports might be sent to clients alongside the month-end proclamation or an assortment letter that gives a nitty gritty record of remarkable things. Along these lines, a records receivable maturing report might be used by the interior as well as outer people.

How Would You Compute Records Receivable aging?

Money due maturing sorts the rundown of open records arranged by their installment status. There are discrete containers for accounts that are current, those that are past due under 30 days, 60 days, etc. In light of the level of records that are beyond 180 days old, an organization can gauge the normal measure of neglected accounts receivables for future benefits.

Why Is Records Receivable aging Significant?

There are two fundamental explanations behind an organization following debt claims maturing. The first is to monitor past due or delinquent records so the organization can keep on chasing after old obligations. These might be offered to accumulations, sought after in court, or basically discounted. The subsequent explanation is with the goal that the organization can ascertain the number of records for which it doesn't anticipate getting installment. Utilizing the remittance strategy, the organization utilizes these evaluations to remember anticipated misfortunes for its fiscal report.

What Is the Commonplace Strategy for Maturing Records?

The maturing technique is utilized to appraise the number of records receivable that can't be gathered. This is generally founded on the matured receivables report, which separates past-due accounts into 30-day containers. Each container is relegated at a rate, in light of the probability of installment. By duplicating all out receivables in each can by the doled out rate, the organization can appraise the normal measure of uncollectable receivables.

Chapter 4

Vo2max by age

VO2max is the number that depicts your cardiorespiratory wellness. A solitary number catches your heart, lungs, circulatory framework, and muscles work all working freely and together. Also, of course, your VO2max is associated with well-being, execution, and life span.

Simultaneously, many individuals find VO2max hard to comprehend. Why? To start with, blending letters, numbers, and truncations together makes things look terrifying and muddled. We should skirt past that issue briefly. On the off chance that you are truly keen on wellness, you will be fine with an entertaining name.

Figuring out VO2max

The main thing to comprehend is that a higher VO2max is by and large better compared to a lower VO2max. A higher VO2max implies that your body is better at taking oxygen from the air and conveying it to your muscles. The more oxygen your muscles can get, the more supplements you can vigorously change into the sub-atomic fuel (ATP) that your muscles use to contract and perform. This is significant on the grounds that your high-impact metabolic pathways are by a long shot your most proficient wellspring of energy for your body.

Higher is superior to lower, got it? Yet, how high will be sufficiently high?

In the event that you are a cutthroat long-distance runner, long-distance runner, cyclist, or cross-country skier, then, at that point, the response is outrageously high. Top perseverance competitors basically commit their lives to work on their VO2max.

For a great many people, in any case, a decent VO2max is grasped concerning by supposed ordinary qualities. These are the scopes of VO2max that scientists have recognized in everyone. Things may then become intriguing at this point. Comparative VO2max results can mean various things for various individuals.

A VO2max of 40 can be brilliant for one individual, really great for another, and quite reasonable for a third. What? The missing setting here is that the primary individual is a 28-year-elderly person, the second is a 42-year-elderly person, and the latter is a 20-year-old male understudy.

Getting a handle on VO2max requires individual settings. That is the genuine test.

VO2max for Men versus Ladies

Variety in actual execution among people generally comes down to contrasts in body creation. Research shows that men commonly have more fit bulk than ladies. Simultaneously, ladies will quite often normally collect more greasy tissue. Where fat stores amass on the body additionally shifts among people. Men will quite often store fat around the storage compartment and mid-region and ladies put away more fat around the hips and thighs.

These normal distinctions in normal body organization are significant for understanding what your own VO2max implies. Muscles use oxygen while fat is essentially put away energy.

Overall, men have higher VO2max values than ladies. Thus, for a man and lady with a similar VO2max, the lady will have a superior wellness level contrasted with her friend bunch.

A top female perseverance competitor will in all likelihood have a lot higher VO2max than the typical male. Nonetheless, she will probably have a lower VO2max contrasted with a top male perseverance competitor.

What's a Decent VO2max for My Age?

Age is generally an intense subject. It's dreadful to ponder, yet our presentation will in general deteriorate as we age. As people, our pinnacle wellness potential is ordinarily around the age of 20. This is valid for all kinds of people.

From that point, wellness normally declines between 5%-20% every 10 years in solid people between the ages of

20 and 65. Cardiorespiratory wellness misfortunes can be overseen through sound way-of-life decisions and ordinary actual work. Past the age of 70 wellness levels decline considerably more rapidly.

A few elements add to progress in years related to wellness declines. One is the way that all our weight or weight will in general increment as we progress in years, however fit bulk diminishes. Another is that our muscles work less productively. This influences the enormous muscles that power our development, yet additionally the actual heart.

As we progress in years, our hearts essentially can't thump as quickly as when we are more youthful. The power with which the heart beats to push oxygenated blood to the muscles additionally diminishes.

The uplifting news about VO2max and maturing is that generally sped-up wellness declines coming about because of stationary ways of life can be turned around. This intends that with appropriate consideration you can work on your wellness and feel more youthful and more fiery all the while.

And My Weight?

Weight the board and wellness points frequently go together and have valid justification Both are great well-being markers and both advantages from sound ways of life. At the point when you get your VO2max from a smartwatch or wellness tracker, the number you see practice researchers call your relative VO2max.

This basically implies that the number you see is how much oxygen you can utilize per kilogram of body weight in a solitary moment. That implies that your body weight is now calculated into the situation.

Factors That Impact VO2 Max Values

The typical stationary male will accomplish a VO2 max of roughly 35 to 40 mL/kg/min. The typical stationary female will score a VO2 max of somewhere in the range of 27 and 30 mL/kg/min. These scores can improve with preparation however might be restricted by specific variables. Among them:

• Age assumes a focal part, with VO2 max scores commonly cresting by age 20 and declining by almost 30% by age 65.

• Orientation likewise contributes to first-class female competitors commonly having higher VO2 max values than their male partners. Be that as it may, when values are changed in view of body size, blood volume, and hemoglobin content, a man's VO2 max will commonly be 20% higher than a lady's.

• Elevation contributes basically in light of the fact that there is less air to consume at higher heights. In that capacity, a competitor will by and large have a 5% diminishing in VO2 max results for each 5,000 feet acquired in altitude.[3]

Higher VO2 max scores are related to specific high-intensity games, explicitly cycling, paddling, distance running, and cross-country skiing. Visit de France champ Miguel Indurain's VO2 max was accounted for at 78 mL/kg/min during the pinnacle of his molding, while cross-country skier Bjørn Dæhlie supposedly accomplished a VO2 max of 96 mL/kg/min.[4]

It is essential to note, nonetheless, that VO2 max values are not innately connected to sports greatness.

While they can surely add to one's prosperity, especially with high-intensity games, different factors ostensibly assume a huge part, including abilities preparation, mental planning, lactate limit preparation, and nourishment.

How the Test Is Performed

VO2 max is ordinarily led in a games execution lab. It is in many cases evaluated, meaning the force is painstakingly aligned and expanded over the long run. Either a treadmill or exercise bike might be utilized.

Before the test, you would be equipped with a facial covering associated with a machine that can break down your respiratory rate and volume close by the centralization of oxygen and carbon dioxide in breathed in and breathed out air. A heart lash would be worn around your chest to gauge your heart rate.[1]

The test for the most part requires somewhere in the range of 10 and 20 minutes. To plan for the test, you would have to:

• Dress in agreeable exercise garments.

• Avoid exercise or preparing 24 hours before the test.

• Stay away from food, liquor, tobacco, and caffeine for no less than three hours prior to testing.

VO2 max is reached when your oxygen utilization stays at a consistent state regardless of an expansion in the responsibility. At this level, the competitor moves from oxygen-consuming digestion to anaerobic digestion. From that point, it is generally not well before muscle exhaustion sets in and powers the competitor to quit working out.

Chapter 5

lifecell muscle anti-aging muscle builder

LifeCell South Ocean Side Skincare sells many beauty care products. Nonetheless, their most popular things arrived in a 4-piece pack comprising of a chemical, hostile to maturing treatment, neck firming serum, and eye treatment.

We'll start by investigating what's inside these items and reach some science-based determinations.

The fixings list(s) are an optimal spot to begin:

LifeCell pH Adjusted Cleaning agent

• Meadowfoam, sunflower, carrot, and Abyssinian seed oils are (like other plant oils) utilized for their calming and antioxidative properties (source).

• Willow bark extrication has been examined as a potential weapon against noticeable indications of skin maturing however not very many investigations exist right now (source).

- Rosemary leaf removal has potential calming activities when applied to the skin (source).

- Cucumber organic product concentrate might decrease skin sebum (oil) content however research demonstrating further advantages is deficient with regards to (source).

- Chamomile removal is utilized for its implied skin-alleviating, mitigating activity (source).

- Green tea extricate is utilized generally in skincare items and, as a rich wellspring of cell reinforcements, could have skin-safeguarding properties (source).

Main concern: LifeCell's chemical depends intensely on plant-inferred fixings, some of which might have antioxidative and calming properties. It is best depicted as a delicate chemical and elements fixings found in numerous other skincare lines.

LifeCell Across the board Skin-Fixing Treatment

Following the chemical, LifeCell prescribes applying their enemy of maturing treatment to "designated regions." Explicitly, the face and neck.

• Dermaxyl is a promoting term for an engineered peptide that might assist with saturating the skin and lessen the presence of lines and kinks. It was developed by a Spanish exploration firm and isn't restrictive to LifeCell (source).

• Hyaluronic corrosive is a staple enemy of maturing fixing that works by hydrating the skin and invigorating the creation of collagen and elastin (source).

• Retinol is another viable and broadly accessible enemy of flaw fixing that can diminish the presence of fine kinks (source).

• L-ascorbic acid is a cancer prevention agent that might advance collagen blend and proposition other skin-defensive advantages (source).

• Ubiquinone has strong cancer prevention agent properties and may, consequently, offer assurance against the harmful impacts of free revolutionaries (source).

• Deanol has possible calming activity and may add to further developed skin solidness (source).

Primary concern: LifeCell's across-the-board enemy of maturing treatment goes to a scope of demonstrated and compelling fixings. Nonetheless, it is actually important that these fixings are accessible in items sold by different brands at lower costs.

LifeCell Neck Firming Serum

• Shiitake mushroom extricate is an uncommon fix on which to base a skin serum. In any case, it might have cancer prevention agents and calming properties (source).

• Dimethicone is most regularly found in hindrance creams since it might improve the skin's safeguards against microorganisms (source).

• Vitamin E is a staple fixing tracked down broadly in skincare items. It is in many cases found in serums implying to work on the presence of stretch imprints as well as scars however more exploration is required before ends on adequacy can be drawn (source).

- Palmitoyl tripeptide-5 is a peptide that might assume a part in further developing skin solidness (source).

This item likewise pairs down on a portion of the fixings we saw before, including ubiquinone (otherwise called Coenzyme Q10), chamomile concentrate, and tea removal.

Primary concern: This. , big neck-firming serum depends on a mix of plant-based fixings combined with peptides, vitamin E, and ubiquinone. Once more, a portion of these fixings might offer unobtrusive advantages however they are not restrictive to LifeCell items.

Under-Eye LifeCell Cooling Treatment

- Shea margarine is a far-reaching saturating fixing that reestablishes skin hindrance capability (source).

- Vitamin D has cell reinforcement properties when applied topically and may offer a level of photoprotection (in any case, it isn't reasonable as a substitute for sunscreen) (source).

- Eyeliss is a reserved term for a peptide/plant removal mix made by the French lab Sederma. They guarantee that the fixing handles eye sacks and puffiness (source).

LifeCell has endeavored to pack an enormous number of fixings into this under-eye treatment, including numerous that are likewise present in prior items in their 'unit'.

It contains ceramides, retinol, plant separates, L-ascorbic acid, a scope of peptides, ubiquinone, and vitamin E.

Main concern: The LifeCell approach ought to be clear enough at this point: pack however many deductively demonstrated enemies of maturing fixings into one item as could be allowed. This technique is noteworthy however it is again worth bringing up that the fixings utilized by LifeCell are all non-selective — importance they're likewise accessible in items sold by different brands.

Scrutinizing LifeCell: Does It Work?

Can we just be real for a minute: promoting in the skincare business is wild. This can pass on many individuals not knowing who to trust and who not to any longer.

Furthermore, it isn't simply an issue of what works and what doesn't. It's an issue of whether premium-evaluated items really work better compared to financial plan choices.

For example, LifeCell gladly expresses that their item is 'roused' by Nobel Prize-winning science. In any case, for all the strong talk, this doesn't pass on us any nearer to understanding how LifeCell items really help the typical client.

Numerous skincare firms stage short preliminaries to give validity to claims that their item is "clinically demonstrated." Notwithstanding, for LifeCell's situation, the PDF containing subtleties of clinical testing shows positively no data on the review conventions. This makes it unthinkable for autonomous specialists to examine the strategies and approaches utilized in the preliminary.

Primary concern: Like most skincare lines, LifeCell backs its items with very intense cases. In any case, they give basically no data on the review convention utilized in their clinical preliminaries.

In any case, we'll utilize this segment of our audit to assess the particular cases made by LifeCell and make a few determinations on whether they're probably going to be valid:

Guarantee: Quickly covers wrinkles with light-reflecting microtechnology

One of LifeCell's more attractive cases is that the item battles against kinks in both the short and long haul. In any case, they appear to be hesitant to carefully describe the situation on precisely which fixing is answerable for the prompt impacts.

In a way that would sound natural to LifeCell, the across-the-board treatment consolidates miniature fillers and "light-reflecting microtechnology" to totally kill the presence of barely recognizable differences.

Our examination concerning this guarantee drove us to focus on one specific fixing: silicon oxide.

Proof on the side of silica arrangements in beauty care products is very frail. Silica nanoparticles may act as an enemy of building up specialists and can improve the 'surface' of skin creams.

Wellbeing controllers in Europe say they have no proof to propose that these nanoparticles are poisonous. In any case, on the grounds that a significant number of these fixings are new to the market, there are still worries that they might convey well-being gambles.

Regardless of whether LifeCell contains effective fixings, they serve just to briefly veil wrinkles (similar to cosmetics). For sure, they don't do anything to definitively neutralize the fundamental reasons for noticeable skin maturing. This repeats a feeling we partook in our Plexaderm survey — one more evidently quick answer for facial lines.

We emphatically urge LifeCell to be more straightforward in examining their 'light-reflecting microtechnology' to open a very close logical examination of the case. These fixings are not extraordinary to LifeCell items so there ought to be no gamble related to free and open sharing.

Ratings

Main concern: LifeCell doesn't uphold their cases of quick flaw-decreasing impacts with tenable science. Regardless of whether the item offers some 'concealing', concealing kinks for a couple of hours never really addresses the hidden reasons for these lines and kinks.

Guarantee: Offers serious firming/Reestablishes 'appearance' of flexibility

Skincare organizations are bosses of cunning wit. Frequently you'll find items promising to 'decrease the presence' of kinks and listing. This can leave customers scratching their heads contemplating whether 'diminished appearance and 'decreased' period are exactly the same thing.

Presently, LifeCell contains a couple of fixings related to unobtrusive enhancements in skin versatility. For example, palmitoyl tripeptide-5, vitamin E, retinol, and deanol.

Unfortunately, notwithstanding, loss of skin versatility and listing are especially hard to turn around seriously. On the off chance that you start utilizing LifeCell with assumptions for sensational improvement, you're probably going to be disheartened.

The American Institute of Dermatology (AAD) says that outcomes from skin-firming creams like this one are "humble, best case scenario, There is a scope of harmless and negligibly intrusive techniques that can help fix and lift the skin.

Be that as it may, these techniques aren't ideal for everybody. For sure, they are a serious and possibly costly game plan.

Ratings

Guarantee: Limits the presence of lines and kinks

LifeCell's scope of items contains basically every staple enemy of maturing fixing available, a large portion of which are shown to be to some degree unassumingly successful in mitigating lines and kinks. For example:

- Emollients
- Peptides

- Nutrients C and E

- Retinol

- Hyaluronic corrosive

- Ceramides

As we saw before, a considerable lot of these fixings offer mitigating and cell reinforcement properties. Hyaluronic corrosive further develops dampness maintenance and retinol and L-ascorbic acid might advance the amalgamation of collagen.

Remember that results will not show up for the time being and will not be all around as 'groundbreaking' as promoting can frequently recommend. These fixings are additionally broadly accessible in other skincare lines.

Ratings

Guarantee: Decreases under-eye packs

Under-eye sacks are normal as we age and are brought about by the debilitating of supporting muscles and liquid amassing. In the meantime, they might be exacerbated by unfortunate rest, sensitivities, and smoking.

Under-eye creams focusing on packs ordinarily center around diminishing the collection of liquid nearby. This can thus assist with diminishing the trademark 'enlarged' appearance of eye packs.

A plant extricates/peptide blend called Eyeliss has all the earmarks of being LifeCell's essential weapon against puffiness. It shows up in the fixings rundown of the cooling under-eye treatment.

Peptides in Eyeliss may likewise add to little upgrades in skin versatility and complexion.

Ratings

What Can and Can't LifeCell Do?

LifeCell incorporates a large group of the notable enemy of maturing fixings. These fixings are accessible in lines sold by other skincare brands.

These items may humbly work on the presence of scarcely discernible differences/wrinkles, listing skin, and staining.

In any case, the outcomes are probably not going to be just about as extraordinary as the when photographs. There are elective arrangements that offer additional enduring outcomes for a portion of the expense. Protection skincare measures incorporate the utilization of sunscreen and adherence to a sound way of life.

Certain corrective strategies can likewise handle indications of skin maturing, however, this course conveys its own arrangement of dangers and can be costly.

Main concern: LifeCell advances its items in every one of the regular ways — strong adverts, shining surveys, and photographs. In any case, these items depend on similarly demonstrated fixings present in a huge number of less expensive lines. Many individuals will see essentially a humble advantage from ordinary use, however, expect nothing extreme!

Is LifeCell Worth The effort?

More or less: **no**.

Surveys on the web will quite often be incredibly captivating. That is, we see shining supports and blistering analysis yet seldom in the middle between.

The objective of our group is to bring a touch of subtlety into the discussion. We research current realities, foster a full picture, then, at that point, draw legit, science-based ends.

In this way, here's the reason we don't suggest LifeCell:

Individuals face significant purchasing choices consistently. For the greater part of us, this is a course of sorting out which item offers the best quality at the most sensible cost.

LifeCell doesn't utilize notable new fixings composed in a clandestine research center! They source a few fixings from global examination firms (who likewise supply different brands) and some from the ordinary channels utilized by all retailers.

For lucidity, we don't really accept that LifeCell is a trick. They are a long-laid-out retailer and they follow a customary (while perhaps not especially momentous) technique in choosing fixings.

Be that as it may, these items neglect to finish the 'esteem assessment'. This is essentially in light of the fact that they don't offer enough 'extra' to legitimize the enormously swelled sticker price. The 4-step unit right now comes in at an incredible $446.

We prescribe getting some margin to peruse one of our purchaser's aides or carrying out your own groundwork to find comparable fixings at a more reasonable cost.

Main concern: LifeCell isn't a trick — notwithstanding, as we would like to think, they really do retail viable and generally accessible enemy of maturing fixings at stunningly expanded costs.

What Do Patients Say?

Client surveys are tormented with dependability issues and it very well may be troublesome in any event, for specialists to tell credible from the phony. This issue positively isn't restricted to LifeCell.

A few retailers endeavor to control online feelings by buying or boosting surveys. In additional grievous cases, brands might try and pay to have contenders flooded with negative surveys.

By the by, we take time in each survey to dissect contemplations from around the web trying to fabricate a wide picture for our perusers.

LifeCell's true site is loaded with shining client surveys. Be that as it may, sentiments somewhere else are very blended.

The organization is recorded on the Better Business Agency (BBB) as 'LifeCell/South Ocean side Skin health management.' It is appraised C-. The normal of eight client audits is just shy of 2/5 stars. Likewise, there are 33 shut grumblings throughout recent years. LifeCell has all the earmarks of being proactive in noting these grievances.

Client audits are tormented with unwavering quality issues and it tends to be troublesome in any event, for specialists to tell the real from the phony. This issue positively isn't restricted to LifeCell.

A few retailers endeavor to control online feelings by buying or boosting surveys. In additional appalling cases, brands might try and pay to have contenders blasted with negative surveys.

By the by, we take time in each survey to examine considerations from around the web trying to fabricate a wide picture for our perusers.

LifeCell's true site is loaded with gleaming client surveys. Nonetheless, sentiments somewhere else are incredibly blended.

The organization is recorded on the Better Business Department (BBB) as 'LifeCell/South Ocean side Healthy skin.' It is appraised C-. The normal of eight client audits is just shy of 2/5 stars. Also, there are 33 shut grievances throughout recent years. LifeCell has all the earmarks of being proactive in noting these grumblings.

Chapter 6

Hunza people

Hunza is a fantasy land and a great deal of legend and reality has been related to it. One thing is valid for all that Hunza individuals are presumably the most agreeable individuals in the district. In the past it was very normal to see individuals crossing 100 and more years this for sure is valid and can in any case be seen somewhat anyway the cutting edge progress has changed a great deal in this valley of longevity.

The Hunza public, or Hunzakuts are individuals who have lived hundreds of years in their own personal confined valleys. They speak Wakhi and Shina. The Wakhi live in the upper piece of Hunza privately called Gojal. Wakhis likewise possess the lining locales of China, Tajikstan, and Afghanistan and furthermore live in the Gizar and Chitral regions of Pakistan. The Shina-talking individuals live in the southern piece of Hunza. They might have come from Chilas, Gilgit, and other Shina-talking areas of Pakistan a long while back.

The Hunzakuts and the area of Hunza have one of the greatest proficiency rates when contrasted with other comparable locales in Pakistan because of the premium of His Highness Karim Aga Khan whom the greater part of the Hunzakuts follow as their spiritual chief.

Neighborhood legend expresses that Hunza might have been related to the lost realm of Shangri La which was referenced in the Novel by James Hilton "The Lost Skyline". Individuals of Hunza are by some prominent for their especially lengthy future, others depict this as a life span fantasy and refer to a future of 53 years for men and 52 for ladies, despite the fact that with an exclusive requirement deviation. The senior Hunza are astounding and nimble and don't need the cutting-edge comforts that matured individuals in the West have, similar to a private home lift or a seat lift. Remaining dynamic and sticking to a customary eating regimen wipes out the requirement for home lifts since they have no trouble crossing steps.

People who call themselves Broshuski, Burusho, or Brusho reside in northern Pakistan's Hunza, Nagar, and

Yasin valleys. There are additionally more than 300 Burusho living in Srinagar, India. They are Muslims in the purest sense. Their language, Burushaski, has not been demonstrated to be connected with some other. They have an East Asian hereditary commitment, proposing that at any rate, a portion of their parentage starts north of the Himalayas

The Hunza and Macedonia

The neighborhood Burusho legend says that individuals of Hunza plunge from the town of Baltit, which had been established by a fighter abandoned by the multitude of Alexander the Incomparable a legend normal to quite a bit of Afghanistan and northern Pakistan. In 1996 an ex-patriate Macedonian etymologist endeavored to show a connection between Burushaski and the cutting-edge, Macedonian language and educated the Hunza regarding the cutting-edge province of the Republic of Macedonia. His proposed semantic association has not been acknowledged by different etymologists, and hereditary proof just backings a Balkan hereditary part in the Afghan Pashtun, not the Burusho. In any case, in 2008 the

Republic of Macedonia coordinated a visit by Hunza Ruler Ghazanfar Ali Khan and Princess Rani Atiqa as relatives of the Alexandran armed force They were welcomed by the State leader Nikola Gruevski and tops of the congregation, however, the resistance excused the visit as populism. This political help of an association with the Hunza matches Greek relations with the adjoining Kalash individuals of Pakistan, who likewise guarantee Alexandran's family line. The issue may hence have more to do with patriotism and the Macedonia naming debate than with the Burusho themselves

Burusho.

The Burushos otherwise called Hunzus, Hunzukuts, and Burushaskis are mountain individuals who live principally in Hunza State and Nagir State in Pakistan. They live in profound valleys and canyons cut by the Hunza Waterway and its feeders, At present, the number of inhabitants in the Burushos surpasses 60,000 individuals. Some live across the Pakistani-Chinese outskirts in the quick-line district of Tibet. Ethnolinguists can't arrange the Burusho language. in any case, it is separated into two lingos that reflect Burusho areas in I-lunza and Nagir. Burusho legend asserts that they plummet from three European troopers left behind when the militaries of Alexander the Incomparable started their retreat from the district. Every one of these troopers established a village. Ba]tir. Ganesh, and Altit-and all Burushos guarantee to drop from the people groups of one of these towns. Burushos live in vigorously braced towns developed 9.000 or 10,000 feet in height and many feet over the Hunza Stream gorge. Most Burushos are resource ranchers who plant their yields in painstakingly gone to terraced fields, Their significant harvests are potatoes. beans, wheat, grain, millet, rye, buckwheat, rice, and various foods grown from the ground. They additionally raise cows, goats, sheep, and chickens, and they keep on hunting to enhance their weight control plans. Burusho society spins around four significant patrilineal factions. every one of them situated in the city of Baltit, and a few minor groups disseminated generally all through the district. The four significant Burusho groups are the Buroongs,* the Diramitings.* and the Baratilangs,"? what's more, the Khurukuts.* notwithstanding the faction framework, Burusho society is partitioned into live classes, including the Thamos. the imperial family; the Uyongko and Akabirting, who lil] most government posts; the Bar, Uncovered. what's more, Sister gatherings, who ranch the land; the Baldakuyos and '1'silgalashos, who are teamsters and transporters for different gatherings; and the Berichos, who are ethnic Indians. The Baldakuyos and Tsilgalashos are the Burushos probably going to find their way across the boundary into China since they assist with shipping items along the Pakistanti-Chinese shipping lanes. Burushos are for all intents and purposes all Muslims of the Ismaili practice. They focus on the Aga Khan as their otherworldly chief. They are more outlandish than different Pakistani Muslims to notice their everyday supplications, quick during Ramadan, and consistently go to the neighborhood mosque.

For a really long time, the l-Iunza Valley in the Karakoram Reach was one of the most disconnected domains of the world. ln 197*8, in any case, Chinese and Pakistani workers finished the development of the Karakoram Thruway, which slice straightforwardly through the Hunza Valley, connecting the area to business shipping lanes between Pakistan and the Individuals

Republic of China. The absolute Burusho population today adds up to just around 60,0lI] individuals. of which two or three hundred inhabit the finish of the Karakoram

Lodgings in Hunza

- Hunza Baltit Hotel Hunza

- Sarai Silk Street Hunza

- Falcons home Duikar Hunza

- Hunza Ridge Lodging Hunza

- Hunza Darbar Inn Hunza

- Hunza Consulate Lodging Hunza

- Hunza View Inn Hunza

- PTDC Inn Hunza

- PTDC Inn Sost Hunza

- Hunza Baltit Hotel Hunza

- Gulmit Mainland Inn Hunza

- PTDC Inn Hunza

- PTDC Inn Sost Hunza
- PTDC Inn Sust Hunza
- Hunza Serena Baltit Motel Hunza
- Gulmit Silk Course Hotel Gulmit
- Passu Touris motel Gulmit
- Passu Traveler hotel Gulmit
- Marcopolo Motel Gulmit
- Marcopolo Motel Gulmit
- Passu Traveler hotel Gulmit
- Gulmit Silk Course Hotel Gulmit

Conclusion

The demands and obligations of their particular practice area will govern how other healthcare professionals engage with older people. The issue of caring for the elderly is incredibly complicated, much like its constituents.

Some subjects will have a direct bearing on them while having a tangential bearing on others. Some people will eventually delve considerably deeper into specific areas due to professional need. But regardless of whether we use the terms senior, elder, older, or geriatric, everyone will be impacted by the need to widen and deepen their understanding of aging and what it means to be an older adult.

There is a demand for changing the way we think about aging and those who are already older, and part of the rationale for doing so is realizing that we are all maturing and will ultimately become older individuals. We need to start thinking of the already old in the same manner if we want to age in a healthy way, be appreciated for what we know and can still do, and be able to make decisions about where to live and what sort of care we want.